Cool Careers in
BIOTECHNOLOGY

Sally Ride
Science

CONTENTS

Bilal

Michelle

Sarkis

Margaret

Gail

Kristala

Shikha

Manolis

Todd

Juan Carlos

Thomas

Amy

What Do You Want to Be?

Is working in biotechnology one of your goals?

The good news is that there are many different paths leading there. The people who work in biotechnology come from many different backgrounds. They include biologists, computer scientists, medical doctors, conservationists, science writers, biochemists, engineers, and many more.

It's never too soon to think about what you want to be. You probably have lots of things that you like to do—maybe you like doing experiments or drawing pictures. Or maybe you like working with numbers or writing stories.

SALLY RIDE
First American Woman in Space

The women and men you're about to meet found their careers by doing what they love. As you read this book and do the activities, think about what you like doing. Then follow your interests, and see where they take you. You just might find your career, too.

Reach for the stars!

Sally K Ride

Michelle was recognized as a Young Innovator by Massachusetts Institute of Technology.

Microbes for Hire?

One day Michelle hopes to put these super microbes she's engineered to work. She hopes they will feed on crop waste—leftovers such as corn husks that are loaded with lignin—and turn them into biofuel. "It's an incredibly complex process that takes many chemical reactions and many genes," says Michelle.

MICHELLE CHANG
University of California, Berkeley

Not for Her?

As a young girl, Michelle Chang often waited and waited while her mom, a scientist, ran experiments in her genetics lab. Michelle decided that she never wanted to work in a lab. That was then. This is now. Michelle not only heads her *own* lab, she's earning awards for her cutting-edge work on making biofuels. "It's exciting," she says. "It's on the frontier of science."

Putting the Pieces Together

Michelle is investigating new ways to make clean, renewable alternatives to fossil fuels—biofuels. The key to the process is breaking down a compound called lignin. Only a few microorganisms can chemically break down this very tough compound. In fact, lignin is so tough that it helps trees stand tall and gives veggies their crunch. Once Michelle grasps how these microorganisms digest the tough stuff, she hopes to bioengineer other microbes so they can digest lots of lignin, too. She'll make copies of the genes for enzymes that help break down lignin and insert them into another microbe's DNA. *Snip, snip, copy, splice*—and she'll have a souped-up microbe.

Lots of lignin—it's the white stuff between these wood cells and inside the cell walls.

A biochemist . . .

studies the chemistry of living things to understand how and why chemical reactions occur. Michelle studies chemical processes in microbes useful in creating new types of biofuels. Other **biochemists**

- determine how medicines work inside the body.
- analyze the nutritional content of foods.
- study how herbicides eliminate weeds from fields.
- identify chemicals that beetles produce for self-defense.

Stack the Deck

As a team, research one kind of biofuel and create a fact card about it. Be sure to include the pros and cons of using it as an alternative to fossil fuels.

Combine your card with other teams' cards. It's a biofuel deck!

Waste to Burn

Today, most biofuels are made from corn—lots of corn. By using crop waste such as corn husks to make biofuels, we can save the corn for people or animals to eat. How much corn can we save?

A bushel of corn weighs 25.4 kilograms (56 pounds), but yields just 10.2 liters (2.7 gallons) of biofuel. Calculate (in kilograms and in pounds, rounded to the nearest tenth) how much corn you need to fill a car's 68-liter (18-gallon) gas tank with biofuel.

Bio Makes it Better

From healthier crops to medical advances and cleaner fuels, biotechnology is improving our lives. Form teams of three, choose one area of biotechnology, and research it. Present a report to your class about what it is and how it makes our lives better. To create the report, each teammate takes a different leadership role—but all team members contribute to each part of the report.

- Lead writer—assembles and edits everyone's text
- Lead illustrator—gathers all photos and creates science drawings
- Lead presenter—organizes the presentation

Check out your answers on page 36.

Device and Conquer

"When you treat patients, you can help them get better one at a time," says Bilal, who grew up watching his father, a doctor, do just that. "But if you develop a device, you can have a much larger impact."

BILAL SHAFI
University of Pennsylvania

One—Or More?

Each year, Bilal Shafi treats hundreds of patients with heart failure. "I used to think, 'Why are we fixing this problem after it's happened? There must be a better way to do this,'" Bilal says. So he combined his training as a doctor and as an engineer, and created a medical device that lets failing hearts heal themselves.

It's a Wrap

Bilal had noticed the heart muscles of patients who had suffered heart attacks were loose and weak and strained to pump enough blood. His new device gives them a needed break. "It's like putting a sock around the heart," Bilal says. The experimental coating works by holding the heart muscles in place and allowing them to heal. It starts as a liquid that the doctor injects. It then quickly turns to gel after wrapping the heart. It's flexible so the heart can beat—but strong enough to support healing muscles. The wrap eventually dissolves—leaving behind a stronger, healthier heart.

Heart wrap

Catheter

Amazing! A liquid gel passes up through the catheter, and then turns into a balloon-like bandage around the healing heart.

A bioengineer . . .

designs, makes, and tests devices used in treating patients. Bilal is developing a coating that helps weakened hearts recover their strength. Other **bioengineers**

- ◘ design replacement hips and other joints.
- ◘ create synthetic skin to treat burn victims.
- ◘ invent artificial bone that can help fractures heal.

About You

As a young boy, Bilal loved to take things apart and then put them back together. What do you like to investigate? Why? Is it hands-on, or not? Write an informative paragraph about it in your About Me Journal.

Big Impact

Approximately 5.7 million people in the U.S. suffer from heart failure—the heart's inability to properly pump blood.

1. If the U.S. population is about 300 million, what percentage of Americans have heart failure?

2. Bilal believes his "heart sock" invention could potentially help one quarter of those with heart failure. How many people is that?

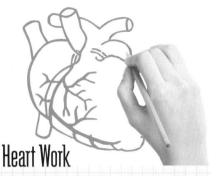

Heart Work

Your heart is hard at work. It beats between 60 to 100 times per minute, pumping oxygen-rich blood to the cells in your body. Pair up with a partner and research the journey blood takes through the chambers of your heart on its way to your cells.

- Draw and label a cross-section of the heart showing how blood flows through the heart. Include the major blood vessels that lead into and out of your heart.
- Use blue to color blood low on oxygen (deoxygenated blood). Which side of the heart pumps blood low on oxygen?
- Use red to color blood full of oxygen (oxygenated blood). Which side of the heart pumps blood full of oxygen?

Use the diagram to teach a younger student how the heart works.

Check out your answers on page 36.

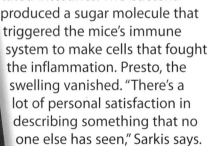

> Sarkis thanks his parents for his strong work ethic. "They taught me there was no substitute for hard work."

SARKIS MAZMANIAN
California Institute of Technology

Curious Child

Sarkis Mazmanian was 12 when he turned the family television inside out. Taking out each part fascinated him—until he tried to put them all back together. Whoops! He still had pieces left over at the end. "My parents had to call a repairman," Sarkis says. "They weren't very happy." What's inside still fascinates Sarkis. These days, though, he's more into what's in *you* and *me* than in the family TV!

*Inner-*resting Science!

Sarkis studies how the trillions of microscopic bacteria that live inside our intestines affect our health. Bacteria usually get a bad rap, even though only a few kinds make us sick. Most are harmless, and some are actually beneficial—maybe even crucial in keeping us healthy, Sarkis says. The bacterium *Bacteroides fragilis* may be a crucial one. Sarkis injected the bacteria into mice with swollen and irritated intestines. The bacteria produced a sugar molecule that triggered the mice's immune system to make cells that fought the inflammation. Presto, the swelling vanished. "There's a lot of personal satisfaction in describing something that no one else has seen," Sarkis says.

Bacteria to the Rescue

Sarkis wants to know if biotechnology can harness the bacterium *B. fragilis* to help people. One day, the microbe could lead to new medicines for Crohn's and other inflammatory diseases. Now that's a gutsy approach to medicine!

Sarkis' research made the cover of this famous journal, plus they used a splashy graphic he helped create!

A biologist . . .

studies living organisms and their interactions with each other and the environment. Sarkis researches how beneficial bacteria found in the human intestine could inspire new biotech medicines. Other **biologists**

- preserve rare plants from tropical forests.
- discover new fish species from the deep ocean.
- isolate genes from bacteria to understand their purpose.
- predict how changing ocean temperatures can affect plankton.

Some Bacteria *Are* Cool

More than 400 types of bacteria live in our digestive systems. Many are beneficial bacteria that break down foods that are difficult to digest.

What are some *other* microorganisms, plants, or animals, where some species are beneficial and some are harmful?

Brainstorm and then research microorganisms with a classmate. Then, as a class, create a Bad Guys and Good Guys Chart.

About Me

What's something you had a good time taking apart or putting together? In your About Me Journal, describe the object, what you did, and what you learned. Then make a labeled diagram of the object.

Some Bacteria *Aren't* Cool

What would happen to milk if you didn't refrigerate it for one week? Break into small groups. Make a prediction, write it on a data sheet, and then do this experiment.

- On a Monday, pour a small amount of milk into each of two plastic cups. Cover them with paper towels.
- Keep one cup at room temperature and the other in a refrigerator.
- Each day, check both cups and log your observations on a data sheet. How does each cup look and smell?
- On Friday, make and record your final observations.

As a class, discuss the findings and conclusions from the experiment.

"Science affects kids' lives in so many ways— no wonder they're fascinated with it."

Don't You Wonder . . . ?

Margaret has been named Arizona's top science teacher several times. "The biggest thing I try to teach is, 'Hey, there is no answer book,'" Margaret says. "No one knows everything about science."

MARGARET WILCH

Tucson High Magnet School

Hands-on, Brains On

How do you teach biotechnology to high schoolers? It's such a fast-moving field that Margaret Wilch advises, "Don't use a textbook—it's instantly outdated!" Students in Margaret's classes learn science by doing science. Working in a high-tech campus laboratory, Margaret's students investigate some cool questions— "Are snack chips made using genetically modified corn?" and "Are soil bacteria in cities more adapted to lead and other heavy metals than soil bacteria found in less polluted rural areas?" Guess what they found? Yes and yes.

Alive and Well

"I try to make biotechnology come alive," Margaret says. That means having her students use the same types of lab equipment and run the same experiments as do real scientists. Margaret's students have extracted the DNA from bacteria, plants, and insects to study their genetic makeup. They have also worked with local biotech scientists whom they have invited to their class. The hands-on experience pays off—Margaret's students regularly take home top awards from science fairs.

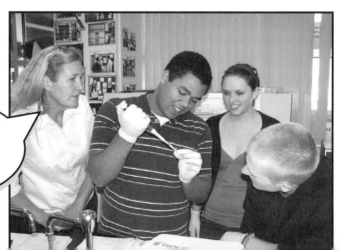

"I'm discovering right alongside my students."

A biotechnology teacher . . .

teaches students how living organisms can be manipulated to make foods, medicines, and other useful products. Margaret teaches her students about biotechnology by having them do biotech experiments. Other **biotechnology teachers** teach students how to

- ☐ create glow-in-the-dark bacteria by splicing in new genes.
- ☐ identify genetically modified vegetables.
- ☐ use a lab technique to make multiple copies of a gene.

Margaret visited the Galapagos Islands and met one of its most famous creatures, a giant tortoise.

Teacher of the Year

Think about what makes Margaret a good teacher. If you taught science, what would make you a good teacher?

- Make a list of qualities and skills you'd need.
- Place an exclamation point next to each one you have.
- Put a check next to those you'd like to work on.

Then write an essay describing what you'd be like as a science teacher. Be sure to include how you would encourage and inspire your students.

New and Improved Veggies

Where do scientists—and science teachers—get their ideas for investigations? Ideas can come from anywhere, even the grocery store. No one likes mushy tomatoes. That led to the development of genetically modified tomatoes that ripen without getting soft.

- Think of a food that you don't like.
- Explain why to a partner.
- Describe what you would change, using genetic engineering, to make the food more appealing.

Lab Laughs

Q. What did the thermometer say to the graduated cylinder?

A. "You may have graduated, but I have many degrees!"

GAIL PHILLIPS
Genentech, Inc.

"School is the ticket to your future. The harder you work, the more places the ticket will take you."

Nose Knows

Growing up, Gail loved science and animals. She was set on becoming a veterinarian—until the itching and sneezing convinced her to pick another career in science.

Did it work? After testing a new cancer drug, Gail adds chemicals that turn dead cells a different color from living cells.

Cancer Killer

Gail Phillips is a lifesaver—her careful lab work led to a powerful new cancer treatment. Examining breast cancer cells under the microscope, Gail noticed something unusual about them. Compared to normal cells, many of the cancerous cells had large numbers of proteins on their surface. Each protein, called a HER-2 receptor, stuck out like an outstretched hand holding a baseball glove. Gail wondered if fitting the right "baseball," or compound, into the glove would stop the cells from growing. She experimented with many antibodies—chemicals that fight infection—to find a fit. Great news—one did! "It was one of the most exciting days of my whole life," Gail says.

Research Ahead

Today that antibody, called Herceptin, is widely used to treat a certain type of breast cancer. "I feel so lucky that I've seen results from my work in the real world."

Herceptin doesn't work in all patients. But that doesn't discourage Gail. It challenges her to work with other researchers to learn more. "When an experiment doesn't work out you have to figure out why," she says. "There is never an end to what you can research."

A cancer researcher . . .

studies the causes of cancer, as well as ways to prevent, diagnose, and treat the disease. Gail researches new breast cancer treatments. Other **cancer researchers**

◻ design techniques to detect cancer early.

◻ study the role diet plays in cancer.

◻ experiment with treating cancer with radiation.

Vaccines for Health

When certain immune cells in your body encounter a microbe, they go on the attack, producing antibodies. Foreign cells destroyed! This is how vaccines work, too.

With a partner, research a disease such as tetanus or measles. Then, create a public service poster about the importance of vaccinations. Include a

- description of the symptoms and possible outcomes of the disease.
- labeled drawing showing how antibodies work to prevent disease.
- slogan emphasizing why vaccinations are important.

Give it a title, and hang your public service poster in your classroom.

Cells by the Billion!

Many cells in the human body are constantly dividing. Overall about 25 million (2.5×10^7) cell divisions take place each second!

At that rate, how many cells would there be after 5 minutes? Express your answer using scientific notation.

HER2 receptor

Normal breast tissue cells (above) produce HER-2 receptors. Cancerous cells (below) produce extra receptors so the cells multiply faster, and so do tumors. Herceptin can block receptors.

Herceptin

Is It 4 U?

As a cancer researcher, Gail

- conducts experiments on cancer cells.
- uses biotech tools to make discoveries.
- communicates with other scientists around the world.

Which parts of being a researcher appeal to you? Write in your About Me Journal explaining what you like, and why.

Check out your answers on page 36.

KRISTALA JONES PRATHER
Massachusetts Institute of Technology

"It's very rewarding to be involved with something new that not everybody else is doing or even knows how to do."

Free Labor

There aren't any pollution-belching smokestacks poking out of the chemical factories Kris Prather builds. Even if there were, you couldn't see them—unless you used a microscope! That's because Kris engineers microbes to become tiny chemical factories. How? First, Kris tinkers with the genes in common bacteria like *E. coli*. Sometimes she inserts genes from other bacteria. This reprograms the microorganisms so they start making useful chemicals like glucaric acid, used to make nylon. It's not just cool, it's clean. Billions of *E. coli* can do the work of a bricks-and-mortar chemical factory—all without the toxic waste a real factory can produce.

Tiny but Mighty

Kris is looking at many different ways of keeping her bioengineered bacteria busy doing useful things. "I love to figure things out," Kris says. One project involves creating an alternative to the phosphates in dish detergents. This chemical can wash into lakes and rivers and make algae grow out of control. By putting bacteria to work, Kris helps keep the environment from going down the drain!

Kris really gets into her work. This photo has been altered so it looks like she's sitting inside a fridge in her lab!

A chemical engineer . . .

designs chemical processes used to manufacture useful products or solve problems, such as pollution. Kris bioengineers bacteria to create chemicals traditionally made in factories. Other **chemical engineers**

◻ monitor the health of the oceans.

◻ create inexpensive ways of manufacturing antibiotic medicines.

◻ make Earth-friendly plastics out of plant compounds.

Phooey to Phosphates

Phosphates from dishwashing detergents can damage our water supply. Imagine you and your business partners have developed a phosphate-free detergent. Work in a small group to

• research phosphates and how they harm the environment.
• create a one-page ad promoting your product.
• come up with a way to inspire your readers to use your phosphate-free detergent.

Ads must make a quick impression. Tell your story visually—with illustrations and a few words that are persuasive and to the point.

Green and Clean?

Sure, store-bought window cleaners do the job. But most are made with toxic chemicals such as glycol ethers. Why not make your own green window cleaner? Just mix equal parts of vinegar and water in a spray bottle. How does it compare to the store-bought cleaner? Experiment and find out!

• As a class, choose *three* places to clean with each window cleaner.
• Make a hypothesis about which will clean better, and why.
• Break into three groups. Each group tests each cleaner in one of the locations.
• On a scale from 1–5, with 5 being best, rate how well each window cleaner worked. Record your results.

As a class, discuss the results. Which cleaner would you choose, and why?

About You

Kris knows that having to try again is more common than having instant success. Write in your About Me Journal about a time you had to try again, and what you learned from it.

SHIKHA ARORA
Life Technologies Corporation

Stick with It

Growing up in India, Shikha Arora liked science, even though some of her friends didn't think it was cool. "They thought it meant wearing a lab coat and big glasses," she laughs. Shikha thought it was so cool she jumped at the chance to move to Michigan after college and study immunology. Now she works at a biotechnology company. She helps to improve a rapid blood test that's used to help match organ donors with needy recipients. Shikha loves that her research helps people, "It can be the difference between life and death."

Matchmaker

The computerized blood test takes just a handful of hours to scan a potential organ recipient's blood for select antibodies. It identifies antibodies that can trigger a patient's immune system into fighting the cells in a donated organ. That means the chip-based test determines whether a patient's immune system might reject an organ before it is transplanted. Shikha continually updates the mix of different antibodies that the test can detect. That ensures it covers as many patients as possible. "We want it to be one-size-fits-all," Shikha says. That way doctors can make the right match every time, with every patient.

"Working for a biotech company gives you the satisfaction of seeing what you're doing directly applied to patients."

Lab Love

In college, Shikha had such a good time doing experiments and making discoveries, she didn't want to leave the laboratory. "They had to practically kick me out!"

Shikha's work continually improves this small super-lab.

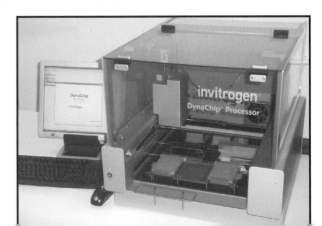

A clinical scientist . . .

helps diagnose and treat diseases and other medical problems, often through the use of laboratory tests. Shikha works on sophisticated blood tests used to match organ donors and transplant recipients. Other **clinical scientists**

☐ develop simpler blood-sugar monitors for diabetics.

☐ create rapid tests to identify patient blood type.

☐ investigate genetic causes of infertility.

Is It 4 U?

What would you like about Shikha's job?

- Finding new ways to save lives
- Working with high-tech lab equipment
- Doing chemistry experiments
- Making medical discoveries

Team up with some classmates and take turns playing the role of Shikha. Choose one part of her job. Then, as Shikha, explain why that's the best part of your job. Go for the Oscar!

Short Supply

Each year, thousands of transplant patients receive new organs. Unfortunately, more people need transplants than there are available donor organs. Sometimes that creates lengthy waiting lists.

Organ	Annual Transplants	Patients on Waiting List
Kidney	16,500	80,500
Liver	6,500	16,500
Heart	2,200	2,800
Lung	1,400	1,900

1. What percentage of patients on the waiting list for each organ receive a transplant each year?
2. Which is the highest percentage? Which is the lowest?
3. How many years would it take for every kidney patient on the waiting list to receive a new kidney?
4. How many years for every heart patient?

Stereotype Types

Shikha's friends thought all scientists wore lab coats and big glasses. After reading about Shikha and other scientists in this book, discuss with a classmate why it's important to not believe stereotypes.

Check out your answers on page 36.

"In life there is no right or wrong answer on what path to follow."

MANOLIS KELLIS
Massachusetts Institute of Technology

Code Breaker

How much fun is it to be a computational biologist? Just take a look at the smile on Manolis Kellis' face. It's almost always there—he's overflowing with energy and passion for his work. Like a code breaker, Manolis uses computers to decipher genomes—the genetic instructions stored inside the cells of every living organism. He turns his knowledge of biology into computational algorithms— step-by-step procedures that solve a problem—for analyzing the genomes of different species.

Of Mice and Men

Manolis has studied the genomes of yeast, fruit flies, and people. By comparing the genomes of many closely related species—for example, 12 different types of fruit flies—he learns how their genomes change over time. He also learns which genes are so important for life that they are preserved unchanged in many species. Manolis and his team at MIT are now comparing the genomes of 29 mammals to better understand ours— the human genome. Manolis first wants to pinpoint regions of the genome associated with certain diseases. If he is successful, that could provide clues about what causes disease—and eventually lead to new cures.

It Was Greek to Him

Manolis was born in Greece and moved to France when he was 12. He didn't know any French, so he had to use a French-to-Greek dictionary to translate his homework problems before solving them. "It was tough," he says, "but it helped me learn how to handle challenges." No kidding! Manolis is one of the top young scientists in the nation—and he has the awards to prove it.

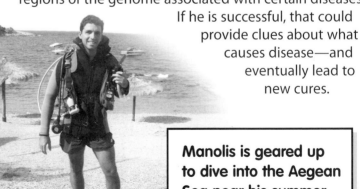

Manolis is geared up to dive into the Aegean Sea near his summer house in Greece.

A computational biologist . . .

uses math and computers to solve problems in biology. Manolis uses computational algorithms to compare the genomes of different species. Other **computational biologists**

☐ virtually screen potential medicines for effectiveness.

☐ extract information from large databases of brain scans.

☐ trace genetic changes that led new species to evolve.

☐ model how flu and other viruses spread in pandemics.

Your Inheritance

Genetic traits are characteristics that are passed down to you by your parents. Create a chart listing these genetic traits.

- Freckles
- Dimples
- Attached earlobes
- Peaked hairline
- Roll tongue
- Hitchhiker's thumb (above)

Record your predictions for which traits are dominant and which are recessive. Then, survey your class. What percentage of students have each trait? *Now* which traits do you think are dominant?

Time Flies!

For more than a century, scientists like Manolis have studied the genetics of fruit flies. What's the big deal about a little fly?

- It takes fruit flies only about 10 days to grow up and reproduce. That means scientists can study multiple generations in a short period of time.
- It takes humans about 25 years to grow up and produce another generation.

How many generations of fruit flies could you raise in 25 years? Round your answer to the nearest whole number.

Is It 4 U?

What parts of Manolis's work would you enjoy most?

- Searching for genetic causes of deadly diseases
- Using math to make discoveries in biology
- Deciphering the genomes of different species

Discuss with a partner which you would enjoy most, and why.

Check out your answers on page 36.

"It's the life on our planet that, as far as we know, makes Earth unlike anywhere else."

De-fense! De-fense!

Ask Todd about the tropics and he'll tell you it's a jungle out there. "Everything is trying to eat everything else—it's a tough environment, with intense competition." Many tropical creatures, like sea sponges, can't flee for safety. Instead, they produce chemicals often toxic to fish and other predators. "Lucky for us," Todd says, "those chemicals often have medicinal properties."

TODD CAPSON
U.S. Department of State

Coral Reef Pharmacy

Todd Capson spent years in tropical countries like Panama. He was bioprospecting—searching for natural sources of potential new treatments for malaria, cancer, and other diseases. Todd organized a team that made one discovery scuba diving off Panama's Coiba Island. They scraped a slimy, deep-red cyanobacterium from a submerged rock. Lab tests later showed it contains a compound that fights tumor cells. "I can't say it will ever become a medicine, but it sure is a promising candidate!"

Save It for Later

"Living libraries." That's what Todd calls tropical ecosystems. "They are full of all sorts of new chemicals," he says. A library worth checking out? You bet—as long as nothing clears the shelves of yet-to-be-discovered organisms *before* bioprospectors arrive to browse. That's why Todd now works full-time preserving and protecting the world's wild places—and things—from overfishing, clear-cut logging, and other destructive human activities.

The largest coral reef in the Eastern Pacific is off Coiba Island.

A conservationist . . .

works to preserve and protect Earth's natural resources. Todd works to save tropical ecosystems that can yield chemical compounds useful for treating malaria and other diseases. Other **conservationists**

- advise farmers on how to stop soil erosion.
- save old-growth redwood forests from being cut down.
- ensure grasslands aren't overgrazed by cattle.
- safeguard drinking water sources from pollution.

Todd says the sea creatures around Coiba, including this purple crab, are "spectacular."

Speak Out

Todd understands why it's important to protect the diversity of life on Earth. But many people do not. Team up to create a radio Public Service Announcement (PSA). Inform people by explaining the importance of biodiversity. Inspire people by focusing on what they can do.

- Write a 30-second script for one or more people to read.
- Audiotape it, and then present it to the class.
- If computers are available, you can create a PSA for TV by adding video clips.

About You

Todd's work takes him around the world, keeping watch on Earth's natural treasures—in the deepest oceans and on the highest mountains. Would you like to travel, scuba dive, or mountain climb as part of your job? What do you enjoy that you'd like to be part of your career? Write your thoughts in your About Me Journal.

To Conserve and Protect

Todd worked to get Coiba National Park listed as a World Heritage Site. That means it will be protected from development. With a partner, research other natural World Heritage Sites. Choose one site and create a science poster about the importance of the site. Be sure to include

- what it is, along with photos or sketches.
- its location on a map.
- why people need to protect it.
- how being listed will protect it.

JUAN CARLOS BREVIS
University of California, Davis

Protein Power

Job Number One for Juan Carlos is to find out which wheat genes produce grains richest in protein. So he extracts the microscopic strands of DNA from the nucleus of wheat cells from the different wheat plants he grows. He then compares the genes to determine which ones might be the plant's "protein genes." That way, Juan Carlos can breed more wheat plants with those same genes. That would be great for Juan Carlos—and for pizza lovers everywhere!

Sowing Solutions

Juan Carlos Brevis adds a high-tech twist to something humans have done for 10,000 years—grow wheat. Juan Carlos sows different types of wheat in the fields surrounding his university. But when he harvests the grains, Juan Carlos doesn't do what's usually done with wheat grains—turn them into flour. He takes the grains to his lab. There, he measures which types produce the most protein—the nutrient essential to build and repair the cells that make up our bodies. Unfortunately, millions of people worldwide don't get enough protein in their diets. One solution is to create new types of protein-rich wheat. "If you can increase the level of protein in wheat, you can have a big impact on humankind," Juan Carlos says. That can lead to more nutritious pastas, breads, crackers, and—yes—pizzas. Next on the menu? Growing onions, such as the ones Juan Carlos is holding above, that are more nutritious.

Nurturing His Nature

Growing up in Chile, Juan Carlos loved running around outside at his grandfather's countryside home. He explored streams, collected insects, and learned plants' names. So when Juan Carlos decided on a career, he chose to stay outside. "I like to see the plants growing," he said.

A plant geneticist . . .

studies the genes of plants in order to isolate certain traits such as plant height or fruit sweetness. Once a plant geneticist finds a desirable trait, she or he uses genetic engineering techniques to ensure that future plants possess the desired traits. Juan Carlos discovers genes in wheat that determine the crop's protein levels. Other **plant geneticists**

- breed drought-tolerant strains of corn.
- genetically modify cotton plants to resist chemical weed-killers.
- raise soybeans that thrive in salty water.
- create rice strains richer in essential nutrients.

Is It 4 U?

What parts of Juan Carlos' job would you enjoy most?

- Growing plants outdoors
- Working in a lab
- Helping improve nutrition
- Designing experiments

Choose one and write a paragraph for your About Me Journal that explains why.

Protein Power

Young people ages 9–13 should eat about 34 grams of protein each day as part of a 2,000-Calorie diet. What are some good sources of protein? A gram of protein contains 4 Calories—do the math.

1. How many protein Calories are in each serving of the foods below?
2. Which food is richest in proteins when measured as a percentage of total Calories?
3. Which food is richest in protein when measured as a percentage of total weight?
4. What food is poorest in protein when measured as a percentage of total Calories?

Food	Serving Size	Total Weight (grams)	Total Calories	Protein Weight (grams)
Plain yogurt	$\frac{1}{4}$ liter	227	139	8
Roasted peanuts	26	28	166	7
Whole wheat bread	Slice	28	69	3
Doughnut	One	47	198	2

Warning—some people are allergic to peanuts.

Wet Wheat

Q. If farmers raise wheat in dry weather, what do they raise in wet weather?

A. Umbrellas!

Check out your answers on page 36.

Plain and Simple

Thomas digs up everything he can about a topic before writing. He reads scientific studies, interviews scientists, and visits biotech labs. Thomas even earned a degree in public health to sharpen his expertise. With all the effort Thomas puts into reporting and writing, it's no wonder he's among the best biotech journalists in America. Keep it up, Thomas!

THOMAS GOETZ

Wired Magazine

Pass the Ketchup

With a nurse for a mom and a doctor for a dad, it's no wonder Thomas Goetz wasn't squeamish growing up. "We talked about blood and surgery and stomachaches at the kitchen table," Thomas says. In eighth grade, Thomas took his interest in medicine beyond just talk—he entered a science fair with an experiment no teacher would ever let him do today. He drew blood from his classmates and tested it for a rare disease called sickle cell anemia. The good news? No one had it—and Thomas took home first prize for explaining how blood tests work.

Reader's Aide

Thomas turned his knack for explaining complicated things into a career—he's now an award-winning journalist. Thomas absorbs ideas from the often complicated world of biotechnology and translates them into stories everyone can understand. That helps his readers see how cutting-edge science affects them. "Helping people to understand these very complicated things is as important as the science itself," Thomas says. Among his stories was an up-close look at spit. That's right—tests on the DNA in saliva can show how our genes put us at risk for ailments like heart disease.

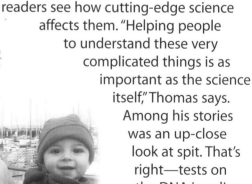

Could Thomas' interest in the future of health care be inspired by his baby Rex?

A science writer . . .

reports and writes about science for print, broadcast, or the Web. Thomas reports on biotechnology for a magazine. Other **science writers**

- ☐ broadcast science news on TV.
- ☐ blog about the latest discoveries.
- ☐ interview scientists for newspapers.

More Spit Tests

Does saliva affect the taste of food? Test it. Get a mix of at least four different dry foods—some salty and some sweet. Then, create a data chart with a *No Saliva* column and a *Saliva* column.

1. Pat your tongue dry with a clean paper towel.
2. Chew a sample of one of the dry foods. How does it taste?
3. Record your observations in the *No Saliva* column.
4. Swallow the sample by drinking some water.
5. Repeat steps 1–4 with the other samples.
6. Repeat the whole experiment except step 1—don't dry your tongue each time.
7. Record these observations in the *Saliva* column.

Compare your data and conclusions with a partner.

Warning—make sure you are not allergic to any of the foods before tasting them.

Junior Journalists

A DNA molecule

Imagine you have been asked to write an article for a teen science magazine about a kind of fruit, vegetable, or grain that has been genetically modified.

- Pick a topic, do some research.
- Pitch the idea to the "magazine editor"— one of your classmates.
- Include *who, what, where, when, why,* and *how.*

Did you convince your "editor" that readers will be interested in the article and learn from it?

Blogging with a Pen

Thomas writes a science blog called The Decision Tree that explores topics of interest to him. If you wrote a science blog, what would you name it? What science topics would you write about?

- Write an "offline" science blog entry on a sheet of paper.
- Have three classmates write comments at the bottom of your entry.
- Respond to the comments.

As a class, discuss the pros and cons of blogging.

AMY WAGERS

Joslin Diabetes Center

The Thrill of Discovery

There's nothing more exciting for Amy Wagers than the sound of success—like a lab researcher running down the hall, yelling, "Amy, Amy, you've got to see this!" When Amy looked through the researcher's microscope, she knew their experiment had worked. She and her lab team had taken a particular type of cell—skeletal muscle stem cells—from the muscle of a healthy mouse. Then they transplanted them into the damaged muscles of a second mouse. Peering into the microscope, Amy saw the transplanted muscle cells had done just as she had hoped. They'd grown to replace the second mouse's damaged muscle cells. They made new muscle! "We were jumping up and down together. It was so exciting!"

Valuable Cells

Amy and her team of researchers study how skeletal muscle stem cells could be coaxed into growing new muscles in people, too. One goal is to isolate stem cells that could replace damaged muscles in people with muscular dystrophy—a disease where the muscles can't repair themselves. The mouse experiment was an early step—but still exciting enough to share with a shout!

"There is an amazing aspect to discovery. You're standing there and saying, 'Gosh, I am the only person in the Universe that knows this right now.'"

"I hope research finds new ways to treat diseases that don't have any treatment."

A stem cell researcher . . .

studies special cells that can develop into many different cell types. Amy uses adult stem cells to regrow muscle cells. Other **stem cell researchers**

- ◻ test new drugs on stem cells instead of people.
- ◻ experiment with regrowing damaged nerves.
- ◻ study restoring damaged pancreatic cells in diabetes patients.

Cell to Cell

Stem cells are extraordinary. They can develop into many different types of cells—including muscle cells and nerve cells. With a group, research two different types of cells that make up the human body. Create a science poster that

- describes the role of each cell type and where it's located.
- compares and contrasts the cell types you researched.
- describes the similarities and differences between the cells you researched and muscle stem cells.

Share your poster with the class. Then brainstorm how muscle stem cells might be useful.

Amy takes to the air! The flying trapeze is one of her hobbies.

Long Division

Stem cells can divide over and over again to make new cells. If a single stem cell divides every 60 minutes to make two new cells, how many total cells will there be after 10 hours? Create a table that shows the number of cells at each hour. Graph your results. What pattern do you notice?

Biotech Tee Hee

Q. Why is math confusing for cells?

A. Because multiplication is the same as division.

Check out your answers on page 36.

About Me

The more you know about yourself, the better you'll be able to plan your future. Start an **About Me Journal** so you can investigate your interests, and scout out your skills and strengths.

Record the date in your journal. Then copy each of the 15 statements below, and write down your responses. Revisit your journal a few times a year to find out how you've changed and grown.

1. *These are things I'd like to do someday.*
   Choose from this list, or create your own.

   - Investigate plant genetics
   - Study the chemistry of microbes
   - Invent medical devices that help people
   - Create biofuels
   - Write about science
   - Explore exotic locations
   - Make food healthier
   - Find new ways to treat diseases
   - Invent environmentally friendly ways to make products
   - Help others understand science
   - Develop new cancer treatments

2. *These would be part of the perfect job.*
   Choose from this list, or create your own.

   - Helping people
   - Working on a team
   - Managing money
   - Designing new products
   - Writing
   - Solving problems
   - Inventing
   - Doing research
   - Building things
   - Working outdoors

3. *These are things that interest me.*
   Here are some of the interests that people in this book had when they were young. They might inspire some ideas for your journal.

   - Dancing
   - Taking things apart
   - Growing vegetables
   - Creating a science fair project
   - Learning about animals
   - Overcoming challenges
   - Learning a new language
   - Making medical breakthroughs
   - Exploring machines, inside and out
   - Investigating nature

4. *These are my favorite subjects in school.*

5. *These are my favorite places to go on field trips.*

6. *These are things I like to investigate in my free time.*

7. *When I work on teams, I like to do this kind of work.*

8. *When I work alone, I like to do this kind of work.*

9. *These are my strengths—in and out of school.*

10. *These things are important to me—in and out of school.*

11. *These are three activities I like to do.*

12. *These are three activities I don't like to do.*

13. *These are three people I admire.*

14. *If I could invite a special guest to school for the day, this is who I'd choose, and why.*

15. *This is my dream career.*

Biotechnology

Which career is 4 U?

What do you need to do to get there? Do some research and ask some questions. Then, take your ideas about your future—plus inspiration from scientists you've read about—and have a blast mapping out your goals.

On paper or poster board, map your plan. Draw three columns labeled **Middle School, High School,** and **College.** Then draw three rows labeled **Classes, Electives,** and **Other Activities.** Now, fill in your future.

Don't hold back—reach for the stars!

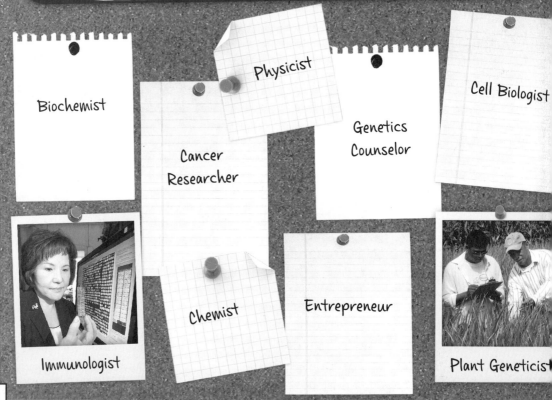

Biochemist

Physicist

Cell Biologist

Cancer Researcher

Genetics Counselor

Chemist

Entrepreneur

Immunologist

Plant Geneticist

Marine
Biotechnologist

Geneticist

Biomedical
Engineer

Biologist

Inventor

Neurobiologist

Journalist

Surgeon

Teacher

Biophysicist

Stem Cell
Biologist

Computational
Biologist

Computer
Scientist

Biostatistician

Chemical
Engineer

Molecular Biologist

Microbiologist

Glossary

antibody (n.) An antibody is a type of protein. The body's immune system produces antibodies when it detects foreign substances, called antigens. Examples of antigens include microorganisms—such as bacteria, fungi, and viruses—and chemicals. Each type of antibody is unique and defends the body against one specific type of antigen. (pp. 14, 15, 18)

bacterium (n.) (*plural* bacteria) A microscopic organism made up of a single cell without a nucleus or other organized cell structures. These organisms occur in three basic shapes—rod, sphere, or spiral. (pp. 10, 11, 12, 13, 16, 17)

biofuel (n.) A fuel made from renewable resources such as plants or municipal wastes. It replaces or reduces the use of fossil fuels such as oil, coal, and natural gas. (pp. 6, 7)

biotechnology (n.) The manipulation, through genetic engineering, of living organisms or their components to produce useful products such as pest-resistant crops, new strains of bacteria, or new medicine. (pp. 7, 10, 12, 13, 18, 26, 27)

chemical reaction (n.) A process where one or more substances are changed into new substances with different properties. (pp. 6, 7)

diabetes (n.) A disease in which there is too much sugar in the blood. A person with diabetes either cannot make enough insulin, the chemical compound that cells need to take in sugar properly, or cannot use it effectively. (p. 29)

DNA or deoxyribonucleic acid (n.) A very large molecule that contains all the information for building and controlling a living organism. It is a double-stranded nucleic acid made of nucleotides—with bases adenine, guanine, cytosine, thymine—and a sugar. DNA is found in the nuclei of all cells, except in bacteria. (pp. 6, 12, 24, 26, 27)

Escherichia coli or *E. coli* (n.) A type of bacteria that lives in your intestines. Most types of *E. coli* are harmless. However, some types can make you sick and cause diarrhea. *E. coli* are used in public health as an indicator of fecal pollution in water or food, and in biotechnology as a research organism. (p. 16)

fossil fuels (n.) Nonrenewable energy resources such as coal, oil, and natural gas that are formed from the compression of plant and animal remains over hundreds of millions of years. (pp. 6, 7)

gene (n.) The unit of heredity, encoded as a specific segment of DNA, in living organisms that determines the characteristics that an offspring inherits from its parent or parents. (pp. 6, 13, 20, 26)

genome (n.) The full set of genes carried by an organism or the range of genes carried by a particular species. The human genome is about 3×10^9 base pairs long. (pp. 20, 21)

immune system (n.) The system that protects the body from infection by microorganisms and disease. It includes the skin and the respiratory, digestive, and circulatory systems. (pp. 10, 18)

immunity (n.) The ability of an organism to resist an infection by harmful microorganisms, and also to fight against cancerous cells. There is some immunity in animals that occurs naturally, for example, provided by the skin barrier and stomach acids, and acquired immunity that occurs as a result of exposure to antigens throughout life. Acquired immunity involves the production of antibodies. (p. 18)

lignin (n.) A chemical compound found in the cell walls of plants, made up of rings of carbon atoms joined in a chain. Lignin provides strength and rigidity to plants. It is difficult to digest so it provides plants with protection from attack by many organisms. (p. 6)

microbe (n.) (also known as a microorganism) A form of life, usually single-celled, that is too small to be seen without a microscope. (pp. 6, 7, 10, 15, 16)

organism (n.) Any living creature, including those made of one cell and those made of many cells (pp. 13, 22)

sickle cell anemia (n.) An inherited blood disorder in which the red blood cells are shaped like a sickle and very fragile. (p. 26)

Index

CHECK OUT YOUR ANSWERS

BIOCHEMIST, page 7

Waste to Burn

Your answers may vary due to rounding. Here, all numbers are rounded to nearest tenth.

169.33 kilograms of corn = 68 ~~liters~~ × $\dfrac{25.4 \text{ kilograms}}{10.2 \text{ ~~liters~~}}$

373.33 pounds of corn = 18 ~~gallons~~ × $\dfrac{56 \text{ pounds}}{2.7 \text{ ~~gallons~~}}$

BIOENGINEER, page 9

Big Impact

1. 1.9 percent has heart failure =

$\dfrac{5.7 \text{ ~~million people~~ have heart failure}}{300 \text{ ~~million people~~ in U.S.}} \times 100$

2. 1.43 million people with heart failure = $5.7 \text{ million} \times \dfrac{1}{4}$

CANCER RESEARCHER, page 15

Cells by the Billion

7.5×10^9 or 7,500,000,000 cells =

$5 \text{ ~~minutes~~} \times \dfrac{25,000,000 \text{ cells}}{1 \text{ ~~second~~}} \times \dfrac{60 \text{ ~~seconds~~}}{1 \text{ ~~minute~~}}$

CLINICAL SCIENTIST, page 19

Short Supply *See Teacher Guide for calculations.*

1. Patients waiting for a
   - kidney = 20.49%.
   - liver = 39.39%.
   - heart = 78.57%.
   - lung = 73.68%.
2. The highest percentage is heart patients. The lowest percentage is kidney patients.
3. It would take about 4.87 years for every patient on the waiting list to receive a kidney transplant.
4. It would take about 1.27 years for every patient on the waiting list to receive a heart transplant.

COMPUTATIONAL BIOLOGIST, page 21

Time Flies

You could raise 913 generations of fruit flies.

913 generations of fruit flies = $25 \text{ ~~years~~} \times \dfrac{1 \text{ generation}}{10 \text{ ~~days~~}} \times \dfrac{365 \text{ ~~days~~}}{1 \text{ ~~year~~}}$

Your Inheritance

- Freckles dominant, no freckles recessive
- Dimples dominant, no dimples recessive
- Detached earlobes dominant, attached earlobes recessive
- Peaked hairline dominant, straight hairline recessive
- Roll tongue dominant, can't roll tongue recessive
- Straight thumb dominant, hitchhiker's thumb recessive

PLANT GENETICIST, page 25

Protein Power *See Teacher Guide for calculations.*

1. One serving of plain yogurt contains 32 protein Calories. One serving of roasted peanuts contains 28 protein Calories. One serving of whole wheat bread contains 12 protein Calories. One doughnut contains 8 protein Calories.
2. Yogurt is richest in protein Calories. Protein makes up 23% of total Calories.
3. Roasted peanuts are the richest in protein by weight. Protein makes up 25% of total weight.
4. Doughnuts are poorest in protein Calories. Protein makes up 4% of total Calories.

STEM CELL RESEARCHER, page 29

Long Division *See Teacher Guide for equation.*

Start with	1 hour	2 hours	3 hours	4 hours	5 hours	6 hours	7 hours	8 hours	9 hours	10 hours
1 cell	1×2 = 2	2×2 = 4	4×2 = 8	8×2 = 16	16	32	64	128	256	512
					$\times 2$ = 32	$\times 2$ = 64	$\times 2$ = 128	$\times 2$ = 256	$\times 2$ = 512	$\times 2$ = 1024 cells

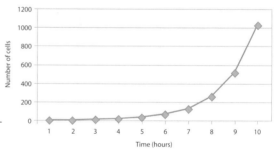

IMAGE CREDITS

© Peter Ginter/Science Faction/Corbis: Cover. Yuri Khristich: pp. 2-5 and 30-31 background, pp. 7-29 banner. Copyright © 2009 UC Regents/Michael Barnes: p. 2 (Chang), p. 6 top. Stanford Biodesign: p. 2 (Shafi), p. 8 top. Greg D'Anna: p. 2 (Wilch), p.12. Melanie Miller: p. 2 (Prather), p. 16 top. Smithsonian Institution: p. 3 (Capson), p. 22 top. Sally Ride Science: p. 4. NASA: p. 5. Bilal Shafi: p. 8 bottom. Reprinted by permission from Macmillan Publishers Ltd./Nature Publishing Group © 2008: p. 10 bottom. Erin Hunter: p. 15. Len Rubenstein: p. 16 bottom. Life Technologies, Inc.: p. 18 bottom. USDA: p. 21 top right. Stephen Kirkpatrick: p. 22 bottom, p. 23. Photo by Michael J. Maloney and courtesy of Joslin Diabetes Center: p. 28 bottom. Clara Lam: p. 30. USDA/Stephen Ausmus: p. 32 bottom left, p. 33 top left. International Rice Research Institute: p. 32 bottom right. Keith Weller: p. 33 bottom left. Ken Hammond: p. 33 top right. U.S. Department of Energy Office of Science/UChicago Argonne LLC: p. 33 bottom right.